THE ESSENTIAL GUIDE TO MARKETING AND SELLING YOUR MOBILE APP

DRIVING SUCCESS IN A COMPETITIVE LANDSCAPE

NADINE NANA TANGPI

PREFACE

Technology touches every corner of our lives, and mobile applications have become one of the most powerful ways for businesses to reach, engage, and serve their audiences. But in today's crowded marketplace, building a great app is only the beginning. Without a thoughtful marketing and sales strategy, even the most innovative product can struggle to gain traction.

The Essential Guide to Marketing and Selling Your Mobile App was created to demystify that process. Nadine has distilled nearly two decades of experience into a practical, approachable resource that breaks down the complexities of app marketing into clear, actionable steps. Whether you're a seasoned founder or a first-time creator, this guide gives you the tools to navigate the mobile landscape with confidence.

What makes this guide special is its accessibility. Nadine takes what can feel like an overwhelming challenge launching and marketing an app and makes it understandable for readers of all backgrounds. You'll find proven strategies, real-world examples, and step-by-step frameworks that you can apply immediately.

As you embark on your app marketing journey, let this book be your trusted companion.

T. Melissa Madian

Founder & Chief Fabulous Officer, TMM Enablement Services Inc.

INTRODUCTION

Why Mobile App Marketing Demands Strategic Precision in B2B SaaS

The mobile app ecosystem has matured into a hyper-competitive, data-rich environment but for B2B SaaS companies, it remains a largely untapped growth lever. While consumer-facing apps dominate headlines, enterprise and vertical SaaS apps are quietly reshaping how businesses onboard, engage, and retain their customers. Whether your app is a companion to your core platform, a mobile-first product, or a critical touchpoint in your customer journey, its success hinges not just on functionality but on strategic go-to-market execution.

As a fractional CMO working with early stage and growth-stage SaaS companies, I've seen the same pattern repeat: teams invest heavily in product development, only to under-resource the marketing and sales engine required to drive adoption. In today's landscape, building a great app is table stakes. The real differentiator is your ability to position it, launch it, and scale it with intention.

The Role of Sales and Marketing in Mobile App Growth

For B2B SaaS, mobile apps are no longer "nice-to-have" they're strategic assets. They can accelerate product-led growth, deepen customer engagement, and unlock new revenue streams. But without a clear sales and marketing strategy, even the most innovative apps risk becoming invisible.

This guide is built on a simple truth: distribution is everything. You need to know how to segment your audience, craft messaging that resonates with decision-makers and end users, and build a repeatable system for acquisition, activation, and retention. That means aligning product, marketing, and sales from day one not as an afterthought, but as a growth imperative.

Who This Guide Is For

This guide is designed for B2B SaaS founders, product marketers, growth leaders, and go-to-market teams who are launching or relaunching a mobile app as part of their customer experience. Whether you're building a mobile-first product, extending your platform to mobile, or using an app to drive adoption and retention, this guide will help you:

- ❖ Clarify your app's strategic role in your GTM motion
- ❖ Design a launch plan that aligns with your sales funnel
- ❖ Optimize discoverability across app stores and owned channels

- ❖ Build a feedback loop between product, marketing, and sales
- ❖ Drive sustainable growth through data-informed iteration

What You'll Learn

We'll walk through the full lifecycle of mobile app marketing from pre-launch positioning to post-launch optimization. You'll learn how to:

- ❖ Conduct market and competitive research that informs your roadmap
- ❖ Define your ideal customer profile (ICP) and user personas
- ❖ Build a brand narrative that resonates across channels
- ❖ Execute a high-impact launch with the right mix of paid, earned, and owned media
- ❖ Align your app's success metrics with broader business goals

This isn't a theoretical playbook. It's a practical, battle-tested guide rooted in real-world experience designed to help you avoid common pitfalls, accelerate adoption, and turn your mobile app into a growth engine.

Let's get started.

TABLE OF CONTENTS

CHAPTER 1
UNDERSTANDING THE MOBILE
APP MARKET

"The mobile app market is a dynamic landscape where understanding user behavior and market trends is crucial for success." **Neil Patel**

WHY MARKET CONTEXT IS YOUR STRATEGIC EDGE

In B2B SaaS, a mobile app is not just a product it's a distribution channel, a retention lever, and often, a competitive differentiator. Whether your app is a mobile-first solution or an extension of your core platform, its success depends on how well it aligns with market needs, user behavior, and business outcomes.

This chapter is your blueprint for understanding the mobile app landscape through a strategic lens. We'll explore how to validate demand, define your ideal customer profile (ICP), analyze competitors, and optimize for discoverability all while keeping your broader go-to-market (GTM) motion in view.

1. Market Research: From Assumptions to Strategic Insight

Too many teams build in a vacuum. Market research is your opportunity to replace assumptions with evidence and align your product roadmap with real-world demand.

Key research dimensions:

❖ Category Trends: Use platforms like CB Insights, PitchBook, and Data.ai to identify macro shifts e.g., the rise of mobile-first field tools, AI-powered productivity apps, or vertical SaaS platforms.

❖ Customer Discovery: Interview both buyers and users. What mobile workflows are broken? What's the cost of inaction? What would make them switch?

❖ Use Case Mapping: Identify the "job to be done" your app solves. Is it about speed, compliance, visibility, or collaboration?

Pro Insight: In B2B, the buyer's motivation (ROI, efficiency, compliance) often differs from the user's (ease, speed, autonomy). Your app must serve both.

2. Define Your ICP and Mobile Use Case

Precision beats generalization. Your mobile app should solve a specific problem for a specific persona not just replicate your desktop experience.

Clarify these dimensions:

Attribute	Example (Field Sales App)
Buyer Persona	VP of Sales Enablement
User Persona	Field Sales Reps
Core Job to Be Done	Capture meeting notes and update CRM on the go
Context of Use	In-transit, between client visits, offline access
Success Metrics	Reduced admin time, faster CRM updates

Tip: Use this clarity to inform your feature set, onboarding flow, and messaging hierarchy.

Bonus Layer: Verticalization

If your app serves multiple industries (e.g., healthcare, construction, logistics), segment your ICPs by vertical. Each will have different compliance needs, workflows, and language.

Tailor your app store listing, sales collateral, and onboarding accordingly.

3. Competitive Intelligence: Learn, Then Leapfrog

Your competitors are your best teachers. Study them not to mimic, but to identify gaps, patterns, and positioning whitespace.

Conduct a focused audit:

- ❖ Feature Set: What's table stakes? What's differentiating?
- ❖ User Reviews: What do users love or hate? What's missing?
- ❖ Positioning Language: How do they frame their value?
- ❖ Pricing Models: Are they bundling mobile access? Charging per seat? Offering freemium?

Case Snapshot: In the B2B scheduling space, Calendly dominates with simplicity and integrations. But tools like Chili Piper differentiate with routing logic and CRM sync. Your edge might be vertical-specific workflows, AI-powered suggestions, or enterprise-grade compliance.

Look for:

❖ Gaps in UX: Are competitors ignoring accessibility, offline use, or multilingual support?

❖ Under-served Segments: Are SMBs priced out? Are enterprise buyers underserved?

❖ Emerging Channels: Are competitors leveraging WhatsApp, Slack, or embedded mobile SDKs?

4. App Store Optimization (ASO): Visibility for the Right Users

ASO is your app's storefront. For B2B apps, discoverability is less about mass appeal and more about clarity, credibility, and conversion.

Keyword Strategy

Use tools like AppTweak, Sensor Tower, and Google Keyword Planner to identify high-intent, low-competition keywords. Focus on terms your ICP would actually search (e.g., "field service CRM," "HIPAA-compliant messaging").

Title & Subtitle

❖ Be clear, not clever. Clarity builds trust.

❖ Include your primary keyword early.

❖ Example: "MedSync: HIPAA-Compliant Messaging for Healthcare Teams"

Description

- ❖ Lead with outcomes, not features.
- ❖ Use bullet points to highlight integrations, compliance, and ROI.
- ❖ Include social proof (e.g., "Trusted by 1,200+ clinics").

Visual Assets

- ❖ **Icon**: Clean, professional, and brand-aligned.
- ❖ **Screenshots**: Show real workflows, not just UI.
- ❖ **Preview Video**: Demonstrate value in under 30 seconds.

5. Beyond ASO: Building Trust and Traction

In B2B, app store visibility is just one piece of the puzzle. You also need to build trust across your owned and earned channels.

Visibility levers:

- ❖ Customer Reviews: Prompt power users to leave reviews after successful outcomes.
- ❖ Release Cadence: Frequent updates signal active development and responsiveness.
- ❖ Sales Enablement: Equip your team with demo videos, one-pagers, and app store links.
- ❖ Partnerships: Co-market with integration partners or industry associations.
- ❖ Thought Leadership: Publish use-case-driven content on LinkedIn, Medium, or your blog to drive organic discovery.

Case Snapshot:

- ❖ **Slack**: Dominates B2B ASO with keywords like "team collaboration" and "remote work." Their screenshots highlight integrations, bots, and enterprise features.

- ❖ **Canva for Teams**: Uses ASO to differentiate its business tier, emphasizing brand kits, team folders, and admin controls.

6. Align Product with Channel: The Product-Channel Fit Test

A common pitfall in SaaS is launching a mobile app without aligning it to the right distribution channel. **Ask:**

- ❖ Is your app designed for self-serve discovery (ASO, SEO)?
- ❖ Or is it a sales-assisted product that needs enablement and demos?
- ❖ Should it be bundled into your core platform or offered as a standalone SKU?

Strategic Tip: Your GTM motion (PLG, sales-led, or hybrid) should shape your app's onboarding, pricing, and positioning.

Chapter Takeaway

Understanding the mobile app market is about more than data it's about strategic empathy. Know your users. Know your buyers. Know your competitors. And above all, know the role your app plays in the broader business ecosystem.

In the next chapter, we'll explore how to build pre-launch momentum from brand positioning to buzz worthy campaigns that prime your market before you even launch.

CHAPTER 2

PRE-LAUNCH STRATEGIES

"Building a strong brand presence and generating excitement before launch can significantly impact the success of your mobile app." **Gary Vaynerchuk**

WHY PRE-LAUNCH STRATEGY IS A GROWTH MULTIPLIER

T he success of a mobile app isn't just about what you build it's about how you frame it, position it, and prime the market before it ever hits the app store. A well-executed pre-launch strategy can compress your time to traction, reduce CAC, and create a groundswell of early adoption.

This chapter outlines how to architect a compelling brand presence, craft a consistent narrative, and generate meaningful buzz all before your app goes live.

1. Build a Strategic Brand Foundation

Your brand is the promise you make and the perception you shape, not only your logo or color palette. In the context of a mobile app, your brand must communicate clarity, credibility, and confidence especially when your app is an extension of a larger SaaS platform.

Define Your Brand Identity and Positioning

Start with your app's strategic role:

- ❖ Is it a standalone product or a companion to your core platform?
- ❖ Is it designed for acquisition, engagement, retention or all three?
- ❖ Who is the buyer, and who is the user?

From there, articulate your:

- ❖ Value Proposition: What problem does your app solve, and for whom?
- ❖ Brand Attributes: Are you authoritative, innovative, human, secure, or mission-driven?
- ❖ **Emotional Resonance**: What do you want users to feel empowered, efficient, in control?

Pro Tip: Use a brand positioning canvas to align your team on who you serve, what you offer, and why it matters.

Create Visual Assets That Signal Trust

Design is not just aesthetic; it's a trust signal. Your app's visual identity should reflect your brand's maturity, clarity, and professionalism.

Key assets to develop:

❖ App Icon: Simple, recognizable, and aligned with your brand system.

❖ Screenshots: Showcase real use cases, not just UI. Highlight workflows, integrations, and outcomes.

❖ Preview Video: A 15–30 second walkthrough that demonstrates value, not just features.

Example: If your app helps field technicians sync job data to a central CRM, show that workflow in action, not a static dashboard.

Craft a Consistent, Conversion-Ready Message

Your messaging should be consistent across every touchpoint from your app store listing to your sales decks to your onboarding emails.

Build a messaging hierarchy:

Layer	Purpose	Example (Compliance App)
Tagline	Core promise	"Compliance, Simplified."
Value Proposition	What it does and why it matters	"Automate audits, reduce risk, and stay compliant."
Feature Highlights	How it delivers	"Real-time alerts, audit trails, and integrations."
Proof Points	Why to trust you	"SOC 2 certified. Trusted by 500+ finance teams."

Pro Tip: Use the same language in your app store listing, website, sales collateral, and onboarding flows. Repetition builds recognition.

2. Generate Pre-Launch Momentum

A successful launch starts long before your app is live. The goal of pre-launch marketing is to build anticipation, gather feedback, and create a pipeline of early adopters who are ready to activate on day one.

Teaser Campaigns and Early Access Programs

Create curiosity and exclusivity by offering a behind-the-scenes look at your app's development.

Tactics to consider:

* ❖ Teaser Videos: Short, high-impact clips showing key features or outcomes.
* ❖ Waitlists: Use tools like Launch List or Type form to collect emails and segment by persona.
* ❖ Beta Access: Offer early access to strategic users (e.g., current customers, partners, or industry influencers).
* ❖ Giveaways & Contests: Incentivize sign-ups with early access, swag, or premium features.

Example: A B2B fintech app might offer early access to CFOs at portfolio companies of their VC partners creating both buzz and credibility.

Activate Social Proof and Influencer Leverage

Even in B2B, people trust people. Use social media and influencer partnerships to amplify your message and build credibility.

Strategies to deploy:

* • LinkedIn Thought Leadership: Share product development updates, behind-the-scenes stories, and customer insights.

- Customer Spotlights: Feature early adopters and their wins.
- Partner Amplification: Co-create content with integration partners or industry associations.
- Micro-Influencers: Collaborate with niche experts who speak to your ICP, think Revamps leaders, product marketers, or compliance consultants.

Pro Tip: Equip your champions with pre-written posts, branded visuals, and shareable assets to make amplification easy.

Collect Feedback and Refine Before You Scale

Your pre-launch phase is the perfect time to test assumptions, validate UX flows, and gather testimonials.

How to structure feedback loops:

- ❖ Closed Beta: Invite a curated group of users to test the app and provide structured feedback.
- ❖ Surveys & Interviews: Use tools like Maze, Type form, or Hotjar to gather qualitative and quantitative insights.
- ❖ Testimonial Capture: Ask beta users for short quotes or video snippets you can use in your launch materials.

Example: "The onboarding was seamless; we had our team up and running in under 10 minutes." Beta user, VP of Operations

Case Studies: Pre-Launch Done Right

Clubhouse (2020)

Clubhouse created massive buzz by launching as invite-only. This exclusivity strategy turned scarcity into demand and fueled viral word-of-mouth a tactic that can be adapted in B2B by offering early access to strategic partners or customers.

Robinhood (2013)

Robinhood's referral program rewarded early users with free stock for inviting friends. This created a viral loop and built a massive waitlist. In B2B, similar mechanics can be applied through partner referrals, customer advocacy programs, or gated feature access.

Chapter Takeaway

Pre-launch is not a warm up, it's your first impression. The strongest B2B SaaS teams treat it like a campaign in itself: strategically branded, tightly messaged, and designed to generate momentum before the first download.

In the next chapter, we'll explore how to execute a high-impact launch, from app store optimization to sales enablement and user onboarding that drives adoption from day one.

CHAPTER 3

LAUNCH EXECUTION

"Effective user acquisition strategies are essential to drive downloads and engagement, ensuring your mobile app reaches its target audience." **Eric Seufert**

WHY LAUNCH IS A STRATEGIC INFLECTION POINT

A mobile app launch is more than a release date it's a coordinated go-to-market moment that can shape your app's trajectory for months to come. In B2B SaaS, where sales cycles are longer and buyer trust is paramount, your launch must do more than generate downloads. It must activate the right users, align with your sales motion, and deliver immediate value.

This chapter walks you through the critical components of a high-impact launch: from app store optimization and creative assets to paid acquisition, partnerships, and social proof.

1. App Store Listing: Your Digital Front Door

Your app store listing is often the first and sometimes only touchpoint a prospect has with your brand. It must communicate value instantly, convert curiosity into action, and reinforce your credibility.

Optimize for Clarity, Relevance, and Conversion

❖ App Title: Use a clear, keyword-rich title that reflects your app's core function. Avoid jargon. Prioritize clarity over cleverness.

  - Example: "Field Sync: Mobile CRM for Sales Teams"

❖ Subtitle (if applicable): Reinforce your value proposition with a concise benefit statement.

  - Example: "Log meetings, sync notes, and close deals on the go."

❖ App Description: Lead with outcomes, not features. Use bullet points to highlight key benefits, integrations, and differentiators.

  - Include keywords naturally to improve search visibility.

❖ App Icon: Design a clean, professional icon that aligns with your brand identity and stands out in a crowded store.

2. Visual Assets That Sell

Visuals are not decoration they're conversion tools. In B2B, they must communicate professionalism, functionality, and trust.

Best Practices for Screenshots and Videos

❖ **Screenshots**:

- Show real workflows (e.g., submitting a report, syncing to CRM).
- Use captions to explain value.
- Sequence them to tell a story: problem → solution → outcome.

❖ **Preview Video**:

- Keep it under 30 seconds.
- Focus on use cases, not UI.
- Use voiceover or captions to reinforce key benefits.

Tip: Include your brand's tone and visual identity across all assets to build consistency and recognition.

3. Activate Social Proof: Reviews, Ratings, and Testimonials

In B2B, credibility is currency. Positive reviews and testimonials can tip the scale for skeptical buyers and end users alike.

Strategies to Encourage and Leverage Feedback

❖ In-App Prompts: Trigger review requests after success moments (e.g., completing a task, syncing data).

- ❖ Follow-Up Emails: Send post-onboarding emails asking for feedback and linking directly to the app store.
- ❖ Respond Publicly: Acknowledge all reviews thank advocates and address concerns with transparency.
- ❖ Repurpose Testimonials: Feature standout quotes in your app store listing, website, and sales decks.

Pro Insight: Early reviews shape perception. Seed your first 10–20 with trusted beta users or internal champions.

4. Paid Acquisition: Precision Over Volume

In B2B, paid campaigns should focus on quality over quantity. Your goal isn't just installs it's activation and retention.

App Install Campaigns

- ❖ **Audience Targeting**:
  - Use firmographic filters (industry, company size, job title) where possible.
- ❖ **Platform Selection**:
  - LinkedIn for enterprise buyers
  - Meta for SMBs and mid-market
  - Google UAC for broader reach
- ❖ **Creative Strategy**:
  - Highlight pain points and outcomes.
  - Use testimonials or stats to build trust.

- Include a strong CTA (e.g., "Start Free Trial," "Sync Your CRM Now").
- ❖ **Campaign Objectives**: Define KPIs beyond CPI such as cost per qualified lead (CPQL), activation rate, or retention at Day 7.

5. App Store Search Ads: Capture High-Intent Users

Search ads place your app in front of users actively looking for solutions. In B2B, this is a powerful way to intercept demand.

How to Win with Search Ads

- ❖ Keyword Strategy: Target long-tail, high-intent terms (e.g., "HIPAA compliant messaging app," "mobile CRM for logistics").
- ❖ Ad Copy: Emphasize outcomes and differentiators.
- ❖ A/B Testing: Continuously test headlines, CTAs, and visuals.
- ❖ Bid Optimization: Monitor performance and adjust bids based on conversion quality, not just volume.

6. Cross-Promotion and Strategic Partnerships

Leverage your ecosystem to amplify reach and credibility.

Partnership Playbook

- ❖ Complementary Apps: Partner with tools your users already use (e.g., Slack, HubSpot, QuickBooks).

- ❖ Co-Marketing Campaigns: Launch joint webinars, blog posts, or bundled offers.
- ❖ Referral Programs: Incentivize partners, customers, or influencers to promote your app.
- ❖ Channel Sales Enablement: Equip resellers or consultants with demo access and marketing collateral.

Pro Tip: Align your app launch with a partner's product release or event for maximum exposure.

7. Launch Case Studies: Lessons from the Field

Strava (2009)

Strava built early traction by embedding social features and gamification into its launch. Their app store listing emphasized community, challenges, and progress tracking turning users into advocates and fueling organic growth.

Calm (2012)

Calm's launch focused on emotional resonance. Their visuals evoked peace, their messaging promised transformation, and their early reviews reinforced credibility. They also invested in ASO and in-app prompts to drive ratings.

Chapter Takeaway

A successful launch is not just about going live it's about going loud, clear, and aligned. In B2B SaaS, your mobile app must launch with the same rigor as your core platform: strategically positioned, visually compelling, and supported by a coordinated acquisition plan.

In the next chapter, we'll explore how to sustain momentum post-launch from onboarding and retention to feedback loops and product-led growth.

CHAPTER 4

POST-LAUNCH ACTIVITIES

"User engagement and retention strategies are the key to creating a loyal user base and maximizing the lifetime value of your mobile app." **Amy Jo Kim**

WHY POST-LAUNCH IS WHERE GROWTH ACTUALLY BEGINS

T he launch of your mobile app is a milestone but it's not the finish line. In B2B SaaS, the real work begins after the initial install. Sustained success depends on your ability to activate users, deliver ongoing value, and build habits that drive retention and expansion.

Post-launch is where your app proves its value not just once, but repeatedly. This chapter explores how to keep users engaged, gather actionable feedback, and continuously improve your app based on real-world usage turning early adopters into long-term advocates.

The 4Rs of Post-Launch Success: A Strategic Framework

To structure your post-launch efforts, I recommend using the 4Rs framework:

1. **Retention:** Keep users coming back by delivering consistent value.

2. **Re-engagement:** Win back inactive users with timely, relevant nudges.

3. **Revenue Expansion:** Identify upsell and cross-sell opportunities based on usage.

4. **Roadmap Feedback:** Use behavioral data and user input to guide product evolution.

This framework ensures your post-launch strategy is not just reactive, but proactive and growth-oriented.

1. User Engagement and Retention Strategies

Push Notifications and In-App Messaging

Used strategically, these tools can nudge users toward value, re-engage dormant accounts, and reinforce key behaviors.

Best practices:

* Segment Your Users: Categorize users by role (admin, end user), lifecycle stage (onboarding, active, dormant), or behavior (power users, at-risk users).

* Personalize Your Messages: Tailor content to user actions, preferences, or milestones.

- **Example**: "You've completed 3 reports this week ready to automate the next one?"

❖ Timing is Key: Use behavioral triggers (e.g., after a task is completed) and respect time zones. Avoid over-messaging, which can lead to fatigue or uninstalls.

Strategic Tip: Use in-app messages for onboarding and feature discovery; reserve push notifications for time-sensitive or high-value nudges.

Personalization and Gamification Techniques

Personalized Recommendations

Use behavioral data to surface the right features, content, or workflows at the right time.

❖ Dynamic Dashboards: Customize the home screen based on user role or usage history.

❖ Smart Suggestions: Recommend integrations, templates, or next steps based on recent activity.

❖ Adaptive Onboarding: Guide users through the most relevant features first, shortening time-to-value.

Gamification Elements

Gamification can be a powerful tool especially when tied to real business outcomes.

- Progress Indicators: Show onboarding completion, usage streaks, or feature adoption milestones.
- Achievements: Reward users for key behaviors (e.g., "First Report Sent," "Team Onboarded").
- Leaderboards: For team-based apps, highlight top performers or most active users (with opt-in privacy settings).

Loyalty Programs and Rewards

Incentivize continued engagement with:

- Usage-based perks (e.g., unlock premium features after X actions)
- Referral bonuses (e.g., invite a colleague, earn credits)
- Milestone rewards (e.g., "100th task completed here's a gift!")

2. Customer Support and Feedback Management

In-App Support Channels

Support is part of the product experience. Make it easy for users to get help without leaving the app.

- Embedded Help Centers: Include searchable FAQs, tooltips, and walkthroughs.
- Live Chat or Chatbots: Offer real-time assistance for common issues or onboarding questions.

- ❖ Escalation Paths: Ensure users can easily escalate to human support when needed.

Promptly Respond to User Feedback

- ❖ Proactive Listening: Prompt users for feedback after key moments (e.g., completing onboarding, using a new feature).
- ❖ Close the Loop: Let users know when their feedback leads to changes it builds trust and loyalty.
- ❖ Feedback Routing: Tag and route feedback to product, marketing, or support teams for faster action.

Pro Insight: Treat feedback as a strategic asset. It's not just about fixing bugs it's about uncovering unmet needs and innovation opportunities.

3. Analyzing App Performance and User Data

Key Performance Indicators (KPIs) Tracking

KPI	What It Measures	Why It Matters
Installs & Activations	Initial reach and onboarding success	Gauges acquisition effectiveness
DAU/WAU/MAU	Daily/Weekly/Monthly Active Users	Indicates engagement and usage consistency
Retention Rate (Day 1/7/30)	User stickiness over time	Predicts long-term value and churn risk

KPI	What It Measures	Why It Matters
Feature Adoption	Usage of key features	Reveals what's working and what's ignored
Conversion Rate	% of users completing key actions	Measures business impact (e.g., trial → paid)
NPS/CSAT	User satisfaction and loyalty	Signals brand health and referral potential

Tip: Align your KPIs with your app's role in the broader customer journey whether it's acquisition, engagement, or expansion.

User Behavior Analysis and Segmentation

Funnel Analysis

Map the user journey from install to activation to expansion. Identify where users drop off and why.

- Example: If 70% of users install but only 20% complete onboarding, your activation flow needs work.

Cohort Analysis

Group users by signup date, industry, or behavior to compare retention and engagement over time.

- Use this to test the impact of new features, onboarding flows, or messaging strategies

Heatmaps and User Flow Analysis

Visualize how users interact with your app. Identify friction points, dead zones, or unexpected behaviors.

- Tools like Mix panel, Amplitude, or Full Story can help you uncover what users do not just what they say.

4. Iterative App Improvements Based on Data Insights

A/B Testing

Test variations of:

- ❖ Onboarding flows
- ❖ Feature placements
- ❖ CTA language
- ❖ Pricing prompts
- ❖ Notification timing

Use statistically significant results to guide product and marketing decisions.

Iterative Updates

- ❖ Ship updates regularly even small ones.
- ❖ Communicate what's new and why it matters.
- ❖ Use changelogs, in-app banners, and email updates to keep users informed.

Monitor User Reviews and Ratings

- ❖ Thank users for positive feedback.
- ❖ Address negative reviews with empathy and solutions.

❖ Track recurring themes to inform roadmap priorities.

5. Cross-Functional Alignment: Product, Marketing, Sales, and Customer Success

Post-launch success is a team sport. Align your internal teams around shared metrics and user insights.

❖ Product: Use feedback and usage data to prioritize roadmap decisions.

❖ Marketing: Build lifecycle campaigns based on user behavior and segment insights.

❖ Sales: Surface product-qualified leads (PQLs) based on in-app activity.

❖ Customer Success: Use engagement data to inform onboarding, QBRs, and renewal strategies.

Strategic Tip: Create a shared dashboard that visualizes key post-launch metrics across departments.

6. Lifecycle Marketing and Behavioral Triggers
Lifecycle Messaging

Design automated campaigns that align with the user journey:

❖ Welcome Series: Guide new users through setup and early wins.

❖ Re-Engagement: Nudge inactive users with personalized "We miss you" messages.

- ❖ Milestone Celebrations: Acknowledge usage anniversaries, achievements, or team growth.
- ❖ Expansion Prompts: Suggest upgrades or add-ons when users hit usage thresholds.

Behavioral Triggers

Use real-time data to trigger relevant messages:

- ❖ "You haven't logged in for 7 days here's what's new."
- ❖ "You've invited 3 teammates unlock your team dashboard."
- ❖ "You're 90% to your monthly goal want help getting across the finish line?"

7. Post-Launch Experimentation Stack

Equip your team with tools to test, learn, and iterate:

Tool Type	Examples	Use Case
Product Analytics	Mix panel, Amplitude, Heap	Track usage, funnels, retention
A/B Testing	Optimizely, Split.io	Test features, flows, messaging
Session Replay	Full Story, Hotjar	Visualize user behavior and friction points
Feedback Collection	Type form, Canny, UserVoice	Capture and prioritize user suggestions
Lifecycle Automation	Customer.io, Braze, Intercom	Trigger messages based on behavior

8. Post-Launch Growth Playbook

Here's a tactical checklist of growth levers to explore:

- ❖ In-app referral programs Encourage users to invite colleagues or peers by offering incentives such as extended trials, feature unlocks, or account credits.
  - *Example*: "Invite a teammate and get 7 extra days of premium access."
- ❖ Usage-based upgrade prompts Trigger upgrade messages when users hit usage thresholds such as exceeding project limits, storage caps, or team size.
  - *Example*: "You've reached your 5-project limit. Upgrade to Pro for unlimited access."
- ❖ Feature gating and progressive unlocks Introduce advanced features gradually, unlocking them as users become more engaged. This builds anticipation and increases perceived value.
  - *Example*: "Unlock advanced analytics after completing your first 3 reports."
- ❖ Product-led sales handoffs Use in-app behavior to surface product-qualified leads (PQLs) to your sales team.
  - *Example*: A user who invites 5+ teammates and uses the app daily may be ready for an enterprise conversation.

- ❖ Community-driven feature requests Let users vote on roadmap items or submit ideas. This builds loyalty and ensures your roadmap reflects real needs.
  - *Example*: Tools like Canny or Upvote can help you collect and prioritize feedback transparently.
- ❖ Customer education loops Launch in-app tutorials, webinars, or certification programs to deepen product knowledge and increase stickiness.
  - *Example*: "Complete our onboarding course and earn a Product Champion badge."
- ❖ Lifecycle-based email campaigns Reinforce in app behavior with email nudges that align with user milestones, inactivity, or feature adoption.
  - *Example*: "You created your first dashboard here's how to automate it."
- ❖ Embedded onboarding checklists Use interactive checklists to guide users through setup and early wins.
  - *Example*: "3 steps to get your team live
  1. Invite teammates,
  2. Connect your CRM,
  3. Set your first goal."
- ❖ Customer success automation Trigger alerts for CSMs when users show signs of churn (e.g., drop in usage) or expansion (e.g., increased team activity).

- *Example*: "User hasn't logged in for 14days time for a check-in."
- ❖ Usage-based pricing nudges If your app uses a usage-based model, show real-time usage dashboards and notify users as they approach thresholds.
  - *Example*: "You've used 85% of your monthly API calls consider upgrading to avoid throttling."

9. Common Post-Launch Pitfalls to Avoid

Even with the best intentions, it's easy to fall into traps that stall growth or erode user trust. Here are some to watch for:

- ❖ Treating launch as the finish line Postlaunch is not maintenance mode it's growth mode.
  - **Instead:** Plan for 90 days of post-launch sprints focused on activation, feedback, and iteration.
- ❖ Over-notifying users Too many push notifications or irrelevant messages can lead to churn.
  - **Instead:** Use behavioral triggers and test frequency caps to balance engagement with respect.
- ❖ Ignoring feedback loops Launching without a system for collecting and acting on feedback leads to blind spots.
  - **Instead:** Build structured feedback channels and close the loop with users.

- ❖ Failing to define success metrics Without clear KPIs, it's impossible to know what's working.
  - • **Instead:** Define and track metrics like activation rate, retention, feature adoption, and NPS.
- ❖ Lack of cross-functional alignment If product, marketing, sales, and success aren't aligned, growth stalls.
  - • **Instead:** Create shared dashboards and rituals (e.g., weekly syncs, QBRs) to stay coordinated.

10. Closing the Loop: Post-Launch as a Growth Engine

Your mobile app is living system, not a static product. Every user interaction, every feedback submission, every drop-off is a signal. The most successful teams don't just monitor these signals they act on them.

By combining behavioral data, user feedback, and experimentation, you can build a post-launch engine that drives:

- ❖ Higher retention
- ❖ Increased customer lifetime value
- ❖ Faster expansion into teams, departments, and organizations
- ❖ Stronger product-market fit over time

This is the essence of product-led growth: letting the product do the heavy lifting of acquisition, activation, and expansion while your team focuses on removing friction, amplifying value, and scaling success.

Chapter Takeaway

Post-launch is not a phase, it's a mindset. It's where your app earns its place in your users' daily workflows and your company's revenue engine. By investing in engagement, feedback, iteration, and cross-functional alignment, you transform your app from a product into a platform for growth.

In the next chapter, we'll explore how to turn that engagement into revenue by selecting the right monetization model, optimizing conversions, and building a sustainable business around your app.

CHAPTER 5

MONETIZATION AND REVENUE GENERATION

"The best monetization strategies align with how customers already experience value." **Wes Bush**

WHY MONETIZATION IS A STRATEGIC DESIGN DECISION

Monetization is not just about pricing it's about aligning value creation with value capture. In the mobile app ecosystem, especially within B2B SaaS, your monetization model must reflect how your app delivers outcomes, integrates into workflows, and supports your broader business model.

Whether your app is a standalone product, a feature extension of your platform, or a customer engagement tool, this chapter will help you evaluate monetization models, optimize revenue per user, and balance profitability with user experience.

1. Choosing the Right Monetization Model

There is no one-size-fits-all approach. Your monetization strategy should be shaped by:

- ❖ Your app's role in the customer journey (acquisition, engagement, retention, expansion)
- ❖ Your target audience's willingness to pay
- ❖ The competitive landscape and pricing norms
- ❖ Your broader business model (PLG, sales-led, hybrid)

Let's explore the most common models and how to apply them strategically.

In-App Purchases and Freemium Models

The freemium model allows users to access a core set of features for free, while offering paid upgrades for advanced functionality, integrations, or services. This model is especially effective for apps that benefit from network effects, virality, or usage-based upsell paths.

Implementation Tips:

- ❖ **Identify Premium Features**:
  - Offer upgrades that deliver clear, measurable value (e.g., advanced analytics, team collaboration, offline access).

- Avoid gating basic functionality focus on features that enhance productivity, scale, or customization.

❖ **Set Competitive Pricing**:

- Benchmark against direct competitors and adjacent tools.
- Use value-based pricing: price according to the ROI your app delivers, not just feature count.
- Consider psychological pricing (e.g., $9.99 vs. $10) and tiered options to anchor value.

❖ **Use Feature Teasing**:

- Show locked features in the UI with tooltips like "Unlock with Pro Plan" to drive curiosity and conversion.
- Offer limited-time trials or usage-based thresholds (e.g., "Free for first 100 tasks").

Pro Tip: Freemium works best when your app has a short time-to-value and a clear upgrade path.

Subscriptions and Tiered Plans

Subscription models are ideal for apps that deliver ongoing value whether through continuous access, regular updates, or recurring services. This model aligns well with SaaS economics and customer expectations.

Key Considerations:

- ❖ **Define Subscription Benefits**:
  - Clearly communicate what users get at each tier.
  - Use benefit-driven language (e.g., "Automate workflows," "Collaborate with unlimited users," "Access premium templates").
- ❖ **Offer Tiered Plans**:
  - Create pricing tiers based on usage (e.g., number of users, storage, integrations) or features (e.g., analytics, support level).
  - Consider a free tier, a core paid tier, and an enterprise tier with custom pricing.
- ❖ **Subscription Periods**:
  - Offer monthly, quarterly, and annual plans.
  - Incentivize longer commitments with discounts (e.g., "Save 20% with annual billing").
  - Use auto-renewal with transparent cancellation policies.
- ❖ **Trial-to-Subscription Flow**:
  - Offer a 7–14-day free trial of premium features.
  - Use in-app prompts and email sequences to convert trial users before expiration.

Example: Zoom's free plan limits meeting duration, nudging users toward paid tiers as usage increases.

Usage-Based and Transactional Models

For apps that facilitate transactions or scale with usage, a pay-as-you-go or commission-based model may be more appropriate.

Use Cases:

❖ **Transactional Apps**:

- Charge a percentage or flat fee per transaction (e.g., Uber, DoorDash).
- Ensure transparency and value alignment users should feel the fee is fair relative to the service.

❖ **Usage-Based Billing**:

- Charge based on API calls, storage, seats, or actions.
- Ideal for infrastructure, analytics, or workflow automation apps.

Strategic Tip: Usage-based pricing aligns revenue with customer success the more value they get, the more they pay.

Advertising and Strategic Partnerships

While less common in B2B, advertising can be a viable revenue stream for apps with high usage volume or niche audiences.

Best Practices:

- ❖ **Ad Formats and Placement**:
  - Use non-intrusive formats like native ads, banners, or rewarded videos.
  - Avoid interrupting core workflows place ads in natural breaks or optional views.
- ❖ **Targeted Advertising**:
  - Use segmentation and behavioral data to serve relevant ads.
  - Consider industry-specific ad networks or direct sponsorships.
- ❖ **Partnerships**:
  - Co-market with complementary tools or platforms.
  - Offer white-labeled versions of your app to partners for a licensing fee.

Example: A project management app might partner with a time-tracking tool to offer bundled features and shared revenue.

2. Implementing Effective Monetization Strategies

Maximize Conversions and ARPU (Average Revenue Per User)

Revenue is a function of both volume and value. Once users are in your app, your goal is to convert them and increase the value of each customer over time.

Tactics to Drive Conversions:

❖ **Clear Call-to-Action (CTA)**:

- Use prominent, benefit-driven CTAs (e.g., "Upgrade to Save Time," "Unlock Advanced Reports").
- Place CTAs at moments of peak motivation after a success, during onboarding, or when a feature is gated.

❖ **Pricing Experiments**:

- Use A/B testing to compare price points, plan structures, and trial lengths.
- Test psychological pricing (e.g., $9.99 vs. $10), bundling, and limited-time offers.

❖ **Upsell Triggers**:

- Prompt upgrades when users hit usage limits (e.g., "You've reached your 5-project limit upgrade to continue").
- Use progress bars, alerts, or modals to visualize value and urgency.

Ad Network Selection and Optimization

If you choose to monetize with ads, selecting the right network and optimizing placement is critical.

Steps to Execute:

❖ **Ad Network Research**:

- Evaluate networks based on CPM, fill rate, ad quality, and targeting capabilities.

- Consider networks like Google Ad Mob, Unity Ads, or niche B2B ad exchanges.

❖ **Placement Optimization**:
- Test different placements (e.g., home screen, post-task, idle screens).
- Monitor CTR, bounce rate, and session duration to avoid cannibalizing engagement.

❖ **Frequency Capping**:
- Limit how often users see ads to avoid fatigue.
- Use frequency caps and user-level controls (e.g., "Remove ads with Pro plan").

3. Balancing Monetization with User Experience

Revenue without retention is short-lived. The most successful apps monetize without compromising usability or trust.

Strategies to Maintain UX Integrity:

❖ **Non-Intrusive Ads**:
- Avoid full-screen interstitials or auto-play videos during core workflows.
- Use native ads that match your app's design and tone.

❖ **Value Exchange Models**:
- Offer rewarded ads (e.g., "Watch a video to unlock this feature").

- Let users opt in to ads in exchange for credits, features, or discounts.

❖ **Transparent Pricing**:
- Be upfront about costs, renewals, and what's included.
- Avoid dark patterns or surprise charges they erode trust and increase churn.

Pro Insight: Monetization should feel like an upgrade, not a tax. Always lead with value.

4. Real-World Monetization Models in Action

Spotify (2008) – Freemium Subscription

Spotify offers a free tier with ads and limited features, and a premium subscription that removes ads, enables offline listening, and unlocks higher-quality audio. Their tiered pricing and student/family plans cater to different segments, driving high conversion and retention.

Candy Crush Saga (2012) – In-App Purchases

Candy Crush monetizes through microtransactions players can buy boosters, extra lives, or level skips. Their success lies in emotional timing (e.g., offering help after failure) and frictionless checkout flows.

Uber (2011) – Commission-Based Model

Uber connects riders and drivers, taking a commission from each fare. Their app monetizes by facilitating transactions, not charging users directly for the app itself a model applicable to marketplaces and service platforms.

Zoom (2013) – Freemium + Tiered Subscriptions

Zoom's free plan offers limited functionality, while paid tiers unlock longer meetings, admin controls, and integrations. Their pricing scales with team size and feature needs, making it ideal for SMBs and enterprises alike.

5. Practical Monetization Checklist

Use this checklist to ensure your monetization strategy is comprehensive, user-aligned, and revenue-optimized:

- ❖ Define your app's role in the customer journey (acquisition, engagement, retention, expansion)
- ❖ Choose a monetization model that aligns with your product's value delivery
- ❖ Identify and test premium features or upgrade paths
- ❖ Design pricing tiers that reflect user segments and usage levels
- ❖ Implement clear, benefit-driven CTAs for upgrades or purchases

- ❖ Use behavioral triggers to prompt upsells at the right time
- ❖ A/B test pricing, messaging, and feature gating strategies
- ❖ Monitor ARPU, conversion rates, churn, and LTV regularly
- ❖ Balance monetization with UX avoid intrusive ads or manipulative patterns
- ❖ Communicate pricing and value transparently across all channels
- ❖ Align monetization with your broader business model and GTM strategy
- ❖ Revisit and refine your monetization model as your app matures

6. Monetization Pitfalls to Avoid

Even well-intentioned monetization strategies can backfire if not carefully implemented. Here are common mistakes and how to avoid them:

- ❖ Over-gating core functionality: If users can't experience value before paying, they won't convert.
- • Instead: Let users experience a meaningful "aha moment" before introducing paywalls.
- ❖ Unclear pricing or hidden fees: Confusion erodes trust and increases churn.
- • Instead: Be transparent about pricing, renewals, and what's included at each tier.

- ❖ One-size-fits-all pricing: Not all users have the same needs or budgets.
- • Instead: Offer flexible plans, usage-based pricing, or modular add-ons.
- ❖ Ignoring monetization until after launch: Retrofitting pricing is harder than designing for it.
- • Instead: Plan monetization alongside product development and GTM strategy.
- ❖ Prioritizing short-term revenue over long-term value: Aggressive monetization can hurt retention.
- • Instead: Focus on delivering value consistently revenue will follow.

7. Monetization as a Growth Lever

App monetization is a strategic tool to validate product–market fit, reinforce customer success, and sustainably fund innovation. Revenue isn't the sole objective, but it is a critical signal that real value is being delivered.

When done right, monetization becomes a growth lever:

- ❖ It helps you segment your audience and prioritize high-value users
- ❖ It creates natural upgrade paths that align with usage and outcomes

❖ It signals value to your market people are more likely to trust and adopt what others are willing to pay for

❖ It funds continued development, support, and innovation

Strategic Insight: Monetization functions as a living system, evolving alongside the product, the users, and the market. Failing to evolve it over time ultimately limits growth.

Chapter Takeaway

Unlocking the revenue potential of your mobile app is a critical step toward building a sustainable, scalable business. Whether you choose freemium, subscriptions, usage-based pricing, or partnerships, your monetization strategy must be intentional, data-informed, and user-aligned.

By designing with value in mind, testing continuously, and balancing revenue goals with user experience, you can create a monetization engine that fuels growth.

In the next chapter, we'll explore how to scale your app's reach and impact through user feedback loops, iterative improvements, and product-led growth strategies that turn satisfied users into your most powerful growth channel.

CHAPTER 6

USER FEEDBACK AND ITERATIVE

IMPROVEMENTS

"Collecting and analyzing user feedback is a vital part of the app development process, allowing you to continually improve and deliver value to your users." **Julie Zhuo**

WHY FEEDBACK IS THE FUEL FOR PRODUCT GROWTH

Your mobile app is never truly "done." It's a living product that evolves through continuous learning. The most successful apps are co-created with users. Feedback is how you close the loop between what you ship and what your users actually need.

This chapter explores how to operationalize feedback collection, analyze insights, prioritize improvements, and build a culture of iteration that drives long-term product-market fit.

1. Building a Feedback-Driven Culture

Before diving into tactics, it's important to establish a mindset: feedback is not a nuisance it's a strategic asset. Every bug report, feature request, or usability complaint is a signal. When you listen, respond, and act, you build trust, improve retention, and accelerate innovation.

Key principles of a feedback-driven culture:
- ❖ Feedback is everyone's responsibility not only product or support.
- ❖ Feedback should be centralized, categorized, and visible across teams.
- ❖ Users should feel heard even if their request isn't implemented.
- ❖ Feedback should inform both short-term fixes and long-term strategy.
- • **Pro Insight:** Treat feedback as a two-way conversation, not a one-time transaction.

2. Collecting and Analyzing User Feedback
In-App Surveys and Micro feedback
Short, contextual surveys are one of the most effective ways to capture user sentiment in real time.

Best practices:
- ❖ Keep it short: Limit to 1–3 questions.

- ❖ Time it right: Trigger surveys after key actions (e.g., completing onboarding, using a new feature).
- ❖ **Ask specific questions**:
  - "What's one thing you wish this app did better?"
  - "How easy was it to complete this task?"
  - "What's your biggest challenge right now?"

Tools to consider: Type form, Survicate, Qualaroo, Appcues, or native SDKs.

Ratings and Reviews

App store reviews are public-facing feedback and a powerful source of insight and social proof.

Tactics to improve quality and quantity:

- ❖ Prompt at the right moment: Ask for reviews after a success event (e.g., completing a task, inviting a teammate).
- ❖ Make it easy: Use native review prompts that don't require leaving the app.
- ❖ Respond publicly: Thank users for positive reviews and address concerns in negative ones.
- ❖ Track sentiment: Use tools like App bot or App Follow to analyze trends and keywords.

Direct Feedback Channels

Create multiple ways for users to share feedback proactively:

- ❖ In-app feedback buttons: "Have an idea? Tell us."
- ❖ Dedicated feedback portals: Use tools like Canny, UserVoice, or Nolt to collect, categorize, and upvote suggestions.
- ❖ **Customer success check-ins**: Train CSMs to capture product feedback during onboarding, QBRs, and support calls.
- ❖ **Slack or Discord communities**: Monitor conversations for pain points, workarounds, and feature requests.

Strategic Tip: Don't collect feedback for the sake of collecting tag it by user segment, plan tier, and lifecycle stage to identify patterns.

3. User Testing and Usability Studies

Quantitative data tells you what users are doing. Qualitative testing tells you why.

In-Person Testing

- ❖ Invite 5–10 users from your ICP to test new features or workflows.
- ❖ Observe their behavior, ask open-ended questions, and note where they struggle.
- ❖ Use prototypes (via Figma, InVision, or clickable mockups) to test ideas before building.

Remote Testing

❖ Use tools like Maze, Playbook UX, or UserTesting.com to run asynchronous tests at scale.

❖ Ask users to complete tasks while narrating their thought process.

❖ Record sessions to identify friction points, confusion, or unmet expectations.

Tip: Run usability tests before and after major releases to validate improvements.

4. Prioritizing Feature Requests and Bug Fixes

Not all feedback is created equal. You need a system to evaluate what to build, fix, or defer.

The RICE Prioritization Framework

Use RICE to score feature requests based on:

❖ **Reach**: How many users will this impact?

❖ **Impact**: How much will it improve their experience?

❖ **Confidence**: How sure are we about the above?

❖ **Effort**: How much time/resources will it take?

$$\text{RICE Score} = \frac{\text{Reach} \times \text{Impact} \times \text{Confidence}}{\text{Effort}}$$

This helps you prioritize high-impact, low-effort improvements.

Bug Triage and Resolution

❖ Log all bugs in a centralized system (e.g., Jira, Linear, Click Up).

❖ Tag by severity: Critical (app-breaking), Major (workflow-blocking), Minor (UI/UX issues).

❖ Set SLAs for response and resolution times.

❖ Communicate fixes in release notes, changelogs, and in-app banners.

Pro Insight: Fixing bugs quickly builds trust. Users are more forgiving of issues when they see you're responsive.

5. Iterative Design and Continuous UX Improvements

Great apps aren't built in one sprint they're shaped over time through iteration.

Embrace User-Centered Design

❖ Involve users early and often in the design process.

❖ Use journey mapping to visualize pain points and opportunities.

❖ Validate new designs with real users before development.

A/B Testing for Optimization

❖ Test variations of onboarding flows, button copy, feature layouts, or pricing prompts.

❖ Use tools like Optimizely, VWO, or Firebase A/B Testing.

- Define clear success metrics (e.g., activation rate, task completion, upgrade rate).
- Run tests long enough to reach statistical significance.

Example: Testing two onboarding flows one with a checklist, one with a video to see which drives faster activation.

6. Closing the Feedback Loop

Collecting feedback is only half the equation. Closing the loop builds trust and drives engagement.

How to close the loop:

- Acknowledge receipt: "Thanks for your suggestion we've logged it for review."
- Update users: "You asked, we listened this feature is now live!"
- Celebrate contributors: Highlight users who inspired new features in release notes or community posts.
- Share your roadmap: Use public roadmaps to show what's planned, in progress, and shipped.

Pro Insight: Transparency builds loyalty. Even when you say "no," explain why.

7. Case Studies: Feedback-Driven Innovation
GitHub (2008)

GitHub has built a culture of feedback through:

- ❖ Public issue tracking and pull requests
- ❖ Community forums and GitHub Discussions
- ❖ Transparent changelogs and release notes

They've launched features like GitHub Actions and Code spaces in direct response to developer needs reinforcing their position as a platform built by and for developers.

Discord (2015)

Discord actively engages its community through:

- ❖ In-app surveys and feedback prompts
- ❖ A public feedback portal with voting
- ❖ Regular updates based on user suggestions

Features like Stage Channels, server folders, and enhanced moderation tools were all born from user feedback helping Discord evolve from a gamer chat app to a full-fledged community platform.

8. Feedback-Driven Product Road mapping

Use feedback to inform not just what you build next but how you shape your long-term vision.

Steps to integrate feedback into your roadmap:

- ❖ Tag feedback by theme (e.g., onboarding, integrations, reporting, mobile UX)

- Quantify demand: How many users are asking for this? What's the revenue impact?
- Balance innovation and iteration: Mix quick wins with strategic bets.
- Review quarterly: Reassess priorities based on new data, market shifts, and customer needs.

Tip: Share your roadmap with internal teams and key customers to align expectations and build buy-in.

Chapter Takeaway

User feedback is not a distraction it's your most powerful growth signal. By embedding feedback into your product development process, you create a virtuous cycle: listen, learn, improve, repeat.

In B2B, where switching costs are high and relationships matter, showing that you listen and act is a competitive advantage. It builds trust, deepens engagement, and ensures your app evolves in lockstep with your users' needs.

In the next chapter, we'll explore how to scale your app's reach and impact through case studies and success stories, learning from companies that have turned feedback, iteration, and strategic execution into breakout growth.

CHAPTER 7

CASE STUDIES AND SUCCESS STORIES

"Examining successful mobile app launches provides valuable insights and inspiration for creating your own path to success in the ever-evolving app market." **Brian Wong**

WHY CASE STUDIES MATTER IN B2B SAAS MOBILE STRATEGY

In the fast-moving world of mobile apps, success leaves clues. Studying how other companies have launched, grown, and monetized their apps can help you avoid common pitfalls, accelerate your go-to-market strategy, and spark ideas you hadn't considered.

But it's not just about copying tactics it's about understanding the principles behind them and adapting those insights to your unique product, audience, and business model.

This chapter offers a structured approach to analyzing successful mobile app case studies, followed by real-world examples across industries. Whether you're launching a new app or optimizing an existing one, these stories will help you think more strategically, act more decisively, and grow more sustainably.

1. How to Analyze a Mobile App Success Story

To extract meaningful insights from a case study, go beyond surface-level metrics. Use this framework to guide your analysis:

A. Product-Market Fit

* ❖ What problem does the app solve?
* ❖ Who is the target user, and how well is the app tailored to their needs?
* ❖ What signals indicate strong product-market fit (e.g., retention, virality, NPS)?

B. Go-to-Market Strategy

* ❖ How did the company generate awareness pre-launch?
* ❖ What channels did they use for acquisition (paid, organic, partnerships, PR)?
* ❖ How did they position the app in the market?

C. Activation and Onboarding

- ❖ How did the app help users reach their first "aha moment"?
- ❖ What onboarding flows, tutorials, or incentives were used?

D. Engagement and Retention

- ❖ What features or tactics kept users coming back?
- ❖ Did they use gamification, personalization, or community?
- ❖ How did they handle feedback and iterate?

E. Monetization

- ❖ What revenue model did they use (freemium, subscription, transactional)?
- ❖ How did they price and package their offerings?
- ❖ What drove upgrades or purchases?

F. Growth Loops

- ❖ Did the app have built-in virality (e.g., referrals, network effects)?
- ❖ How did they leverage user behavior to drive organic growth?

2. Real-World Case Studies: Lessons in Action

Airbnb (2009) Community-Driven Growth

Category: Marketplace / Travel

Model: Transactional + Platform Fees

Key Takeaways:

- ❖ Pre-Launch: Airbnb built early traction by manually onboarding hosts and photographing listings themselves a scrappy, high-touch approach to building supply.
- ❖ Launch Strategy: They leveraged Craigslist to cross-post listings, tapping into an existing user base.
- ❖ Trust as a Feature: Reviews, host profiles, and secure payments were core to the product — not just add-ons.
- ❖ Mobile Extension: Their app made it easy for travelers to book on the go and for hosts to manage listings in real time.

Insight: If your app facilitates transactions or services, focus on trust, liquidity, and seamless mobile workflows. Consider how your app can reduce friction in real-world interactions.

Headspace (2010) Personalization and Partnerships

Category: Health & Wellness

Model: Freemium + Subscription

Key Takeaways:

- ❖ Content Personalization: Headspace used onboarding quizzes to recommend meditation tracks tailored to user goals (e.g., stress, sleep, focus).

- ❖ Partnerships: They partnered with companies like Nike and airlines to expand reach and credibility.
- ❖ Retention Strategy: Daily reminders, streaks, and progress tracking helped build habits.
- ❖ Enterprise Expansion: Headspace for Work brought mindfulness into corporate wellness programs.

Insight: Personalization isn't just for consumers. In B2B, tailoring onboarding, content, and notifications to user roles and goals can dramatically improve engagement and retention.

Duolingo (2012) Gamification at Scale
Category: EdTech
Model: Freemium + Ads + Subscription

Key Takeaways:
- ❖ **Gamified UX**: Duolingo turned learning into a game with XP points, streaks, leaderboards, and daily goals.
- ❖ **Data-Driven Iteration**: They ran thousands of A/B tests to optimize everything from button copy to lesson structure.
- ❖ **Freemium Funnel**: The free tier was generous, but ads and time gates nudged users toward paid plans.
- ❖ **Community and Crowdsourcing**: Users contributed to course creation, expanding content at scale.

Insight: Gamification can drive engagement in professional tools too think onboarding checklists, usage milestones, or team leaderboards for productivity.

Slack (2013) Product-Led Growth in B2B

Category: Communication / Collaboration
Model: Freemium + Tiered Subscription

Key Takeaways:

* **PLG Motion**: Slack grew bottom-up individual teams adopted it, then expanded across organizations.
* **Onboarding Excellence**: Their "Slack Bot" guided users through setup with a friendly, conversational tone.
* **Viral Loops**: Every new user invited others, creating exponential growth.
* **Enterprise Expansion**: Slack added admin controls, compliance features, and integrations to support larger teams.

Insight: If your app supports team workflows, design for virality and internal expansion. Make it easy for users to invite others and demonstrate value quickly.

Loom (2016) Feedback-Driven Iteration

Category: Video Communication

Model: Freemium + Subscription

Key Takeaways:

❖ **Rapid Iteration**: Loom launched with a simple MVP and improved based on user feedback.

❖ **Community-Led Growth**: They built a passionate user base that shared Loom videos publicly, driving organic adoption.

❖ **PLG + Sales Hybrid**: As usage grew, Loom layered in sales for enterprise accounts.

❖ **Product-Led Support**: In-app feedback tools and usage analytics informed roadmap decisions.

Insight: Start simple, listen closely, and let your users shape the product. Feedback loops are your fastest path to product-market fit.

3. Applying Case Study Insights to Your App

Identify Patterns, Not Just Tactics

Look for recurring themes across success stories:

❖ Clear value proposition

❖ Fast time-to-value

❖ Seamless onboarding

- ❖ Built-in growth loops
- ❖ Feedback-driven iteration
- ❖ Strategic monetization alignment

Adapt, Don't Copy

Every app operates in a unique context. What worked for Duolingo may not work for a compliance automation tool. Instead of copying tactics, ask:

- ❖ What's the principle behind this strategy?
- ❖ How can I adapt it to my audience, product, and market?
- ❖ What constraints or opportunities are unique to my space?

Innovate on the Edges

Use case studies as a springboard not a blueprint. Combine proven strategies with your own insights, experiments, and creativity.

- ❖ What's missing in your category?
- ❖ How can you surprise and delight users in ways others haven't?
- ❖ What can you do better, faster, or more authentically?

Chapter Takeaway

Case studies are more than stories they're strategic mirrors. They reflect what's possible, what's repeatable, and what's worth reimagining. By studying successful mobile apps through a structured lens, you gain more than inspiration you gain a toolkit for execution.

Whether you're launching a new app or scaling an existing one, let these stories guide your thinking, sharpen your strategy, and remind you that success is not a formula it's a process of learning, adapting, and delivering value consistently.

In the next chapter, we'll explore how to secure funding for your mobile app from bootstrapping and angel rounds to venture capital and strategic partnerships.

CHAPTER 8

SECURING FUNDING FOR YOUR MOBILE APP

Once you have real users and real data, it is time to think seriously about raising investment." **George Berkowski**

Securing funding is one of the most pivotal steps in transforming your mobile app from concept to scalable product. Whether you're a solo founder or leading a venture-backed startup, the funding landscape has evolved.

Investors now prioritize traction, capital efficiency, and AI-readiness over raw ideas. This chapter explores both traditional and emerging funding paths from bootstrapping and angel networks to micro-VCs, syndicates, and revenue-based financing and offers practical guidance on how to pitch, position, and align your app with the right capital at the right time.

1. Traditional Funding Options (Still Relevant, But Evolving)

❖ **Bootstrapping**

Still the most common starting point. Use personal savings, consulting income, or early revenue to fund MVP development.

- **Pro Tip**: Use no-code tools like **Flutter Flow** or **Bravo Studio** to reduce dev costs by up to 40% before raising capital.

❖ **Friends & Family**

Approach with professionalism. Provide a clear business plan, risk disclosures, and formal agreements.

- **Pro Tip**: Use convertible notes or SAFE agreements to avoid early valuation debates.

❖ **Pitch Competitions**

Great for early visibility, feedback, and non-dilutive capital.

- **Where to Look**: TechCrunch Startup Battlefield, Y Combinator's Startup School, and local accelerators.

❖ **Bank Loans**

Still viable for founders with strong credit and collateral.

Pro Tip: Some fintech lenders now offer app-specific loans based on projected ARR or app store performance.

2. Angel Investors and Syndicates

Angel Investors

Angels invest $10K–$500K in early-stage startups. They value traction, team strength, and a clear path to monetization. Angels now expect at least 30 days of MVP validation and 20–30% MoM user growth.

Micro-Angels and Syndicates

Platforms like **AngelList**, **Seed Invest**, and **Republic** allow you to pitch to curated groups of micro-angels.

- ❖ Why It Works: Faster decisions, smaller checks, and less red tape than traditional VC.
- ❖ Pitch Tip: Lead with ROI. "Your $50K secures 20% of our $500K ARR runway."

3. Crowdfunding and Community Capital

Reward-Based Crowdfunding

Platforms like **Kickstarter** and **Indiegogo** are ideal for B2C apps with strong visual appeal or early prototypes.

Best Practices:

❖ Offer early access, branded perks, or lifetime discounts.

❖ Use video storytelling to build emotional connection.

❖ Promote heavily before launch 80% of success comes from pre-launch momentum.

Equity Crowdfunding

Platforms like Wefunder, StartEngine, and Republic allow you to raise capital from retail investors.

New Trend: Equity crowdfunding is gaining traction for mission-driven or niche apps with strong communities.

4. Venture Capital (VC) High Stakes, High Standards

VCs fund high-growth startups with large markets and proven traction.

Reality Check: Less than 1% of startups secure VC funding.

What VCs Want:

❖ LTV/CAC ratio > 3

❖ Clear path to $10M+ ARR

❖ AI-readiness and defensible tech

❖ Team with domain expertise and execution history

Avoid This: Pitching pre-revenue apps to VCs — 89% are auto-rejected.

Emerging Funding Models

Revenue-Based Financing (RBF)

Platforms like **Pipe**, **Capchase**, and **Founderpath** offer non-dilutive capital based on recurring revenue.

Ideal For: SaaS apps with predictable MRR.

How It Works: You repay a percentage of monthly revenue until the loan is paid off.

Grants and Innovation Funds

Government and private grants are available for apps in healthtech, climate, education, or social impact.

Pro Tip: Only apply if your app aligns with grant priorities 92% of misaligned applications are rejected.

Corporate Venture and Strategic Partnerships

Large tech companies and industry incumbents are investing in apps that complement their ecosystems.

Examples:

- ❖ Salesforce Ventures (B2B SaaS)
- ❖ Google for Startups
- ❖ AWS Activate for infrastructure credits

What Investors Look For

Criteria	What It Means
Market Potential	Large, growing TAM with clear pain points and underserved segments
Traction	Real usage data, retention metrics, and revenue (or strong leading indicators)
Differentiation	10x better, faster, or cheaper than alternatives; ideally AI-enhanced
Team Credibility	Founders with domain expertise, execution history, or technical depth
Capital Efficiency	Lean burn, clear milestones, and a path to profitability or scale

7. Pitching for the Right Stage

Pitch Type	When to Use	Format
Elevator Pitch	Networking events, cold outreach	60 seconds, 1–2 sentences
Teaser / Executive Summary	Email intros, investor platforms	1–2 pages
Overview Deck	First meetings, demo days	8–12 slides
Detailed Pitch	Second meetings, due diligence	15–20 slides + appendix

Pro Tip: Use AI tools like **Beautiful.ai**, **Tome**, or **Gamma** to create dynamic, data-rich decks that stand out.

Chapter Takeaway

Funding your mobile app in this new competitive market requires more than a great idea it demands traction, clarity, and alignment with the right capital source. Whether you bootstrap, crowdfund, pitch angels, or pursue venture capital, your success depends on matching your app's stage with investor expectations and telling a story that resonates.

In the next chapter, we'll explore how to scale your app's reach and revenue through advanced growth strategies from product-led expansion to channel partnerships and internationalization.

Would you like me to build a pitch deck outline tailored to your app's stage and funding goals?

CHAPTER 9

SCALING YOUR MOBILE APP: GROWTH STRATEGIES BEYOND LAUNCH

"Growth is never by mere chance; it is the result of forces working together." **James Cash Penney**

WHY SCALING IS A DISCIPLINE, NOT A SPRINT

L aunching your mobile app is a milestone but scaling it is a discipline. It requires more than just marketing spend or feature releases. True growth comes from aligning product, marketing, sales, and customer success around a shared goal: delivering compounding value to your users.

In this chapter, we'll explore how to scale your mobile app through product-led growth (PLG), channel partnerships, international expansion, and data-driven experimentation.

Whether you're bootstrapped or venture-backed, these strategies will help you grow with intention not just velocity.

1. Product-Led Growth (PLG): Let the App Sell Itself

PLG is a go-to-market strategy where the product is the primary driver of acquisition, activation, and expansion. It's especially powerful for mobile apps, where users expect to try before they buy.

Key PLG Levers:

- ❖ Self-Serve Onboarding: Design intuitive flows that guide users to their first "aha moment" without human intervention.
- ❖ Usage-Based Triggers: Prompt upgrades or referrals based on behavior (e.g., "You've hit your usage limit unlock more with Pro").
- ❖ In-App Expansion: Make it easy for users to invite teammates, unlock features, or upgrade plans directly in the app.
- ❖ Product-Qualified Leads (PQLs): Use in-app behavior to identify high-intent users and route them to sales.

Pro Tip: Track activation metrics like time-to-value, feature adoption, and Day 7 retention to optimize your PLG funnel.

2. Channel Partnerships: Multiply Your Reach

Strategic partnerships can accelerate growth by tapping into existing audiences, distribution channels, or ecosystems.

Types of Partnerships:

❖ Integration Partners: Build native integrations with popular tools (e.g., Slack, HubSpot, Salesforce) to increase stickiness and discoverability.

❖ Reseller or Affiliate Programs: Empower agencies, consultants, or influencers to promote your app in exchange for commissions.

❖ Platform Ecosystems: List your app on marketplaces like the Apple App Store, Google Play, Shopify App Store, or Microsoft AppSource.

❖ Co-Marketing Campaigns: Collaborate on webinars, content, or bundled offers with complementary products.

Example: A project management app partnering with a time-tracking tool to offer a bundled productivity suite.

3. International Expansion: Think Global, Act Local

Mobile apps are inherently global but scaling internationally requires more than just flipping a language switch.

Steps to Expand Globally:

- ❖ Market Prioritization: Use data to identify high-potential regions based on app store traffic, search volume, and competitor presence.
- ❖ Localization: Translate not just language, but cultural context, pricing, and support.
- ❖ Local Payment Methods: Support region-specific payment gateways (e.g., UPI in India, IDEAL in the Netherlands).
- ❖ Regulatory Compliance: Ensure GDPR, data residency, and local tax compliance.
- ❖ Local Partnerships: Collaborate with regional influencers, resellers, or media outlets.

Strategic Insight: Start with one region, test your localization playbook, then scale to adjacent markets.

4. Advanced Growth Tactics

Lifecycle Marketing

Use email, push notifications, and in-app messaging to guide users through their journey:

- ❖ Onboarding Sequences
- ❖ Reactivation Campaigns
- ❖ Feature Discovery Nudges
- ❖ Milestone Celebrations
- ❖ Upgrade Prompts

Community-Led Growth

Build a user community that drives engagement, support, and advocacy:

❖ Launch a private Slack, Discord, or Circle community

❖ Host AMAs, webinars, or product roundtables

❖ Empower superusers as ambassadors or beta testers

Content and SEO

Create high-value content that attracts and educates your ideal users:

❖ Publish use-case-driven blog posts, case studies, and comparison pages

❖ Optimize for app store search (ASO) and web SEO

❖ Repurpose content into videos, carousels, and email sequences

5. Growth Metrics That Matter

Metric	Why It Matters
Activation Rate	% of users who reach first value moment
Retention Rate (Day 1/7/30)	Measures stickiness and long-term value
Referral Rate	Indicates virality and user satisfaction
Expansion Revenue	Upsells, cross-sells, and team growth
LTV/CAC Ratio	Measures efficiency of customer acquisition
North Star Metric	Your app's core value delivery (e.g., tasks completed, messages sent)

Tip: Choose one North Star Metric and align your team around improving it.

6. Scaling Pitfalls to Avoid

❖ Scaling before product-market fit

❖ Over-relying on paid acquisition without retention

❖ Ignoring onboarding friction

❖ Expanding internationally without localization

❖ Chasing vanity metrics over sustainable growth

- **Instead:** Focus on compounding value, user-led expansion, and data-informed decisions.

Chapter Takeaway

Scaling your mobile app isn't about doing more it's about doing what works, better. By aligning your product, growth, and customer success strategies around user value, you create a flywheel that drives sustainable, scalable growth.

In the next chapter, we'll explore how to measure success from defining your North Star Metric to building a growth dashboard that keeps your team focused and accountable.

If you'd like, I can also help you design a sample growth dashboard or onboarding funnel tailored to your app's goals.

CHAPTER 10

MEASURING SUCCESS: METRICS, DASHBOARDS, AND DATA-DRIVEN DECISIONS

"What gets measured gets managed." **Peter Drucker**

WHY METRICS MATTER MORE THAN EVER

In the mobile app world, intuition can spark innovation, but only data sustains it. Whether you're optimizing onboarding, pitching investors, or scaling internationally, your ability to measure what matters is what separates momentum from mediocrity.

This chapter will help you define your app's North Star Metric, select the right supporting KPIs, and build a growth dashboard that keeps your team aligned, accountable, and agile.

1. Define Your North Star Metric (NSM)

Your North Star Metric is the single most important metric that captures the core value your app delivers to users. It should reflect both user success and business growth.

Characteristics of a Strong NSM:

- ❖ Tied to user value (not vanity)
- ❖ Measurable and trackable
- ❖ Correlated with retention and revenue
- ❖ Easy to understand across teams

App Type	Example NSM
Project Management	Tasks completed per active user
Meditation App	Sessions completed per week
Language Learning	Lessons completed per user
B2B CRM	Deals closed per sales rep
Video Messaging	Videos recorded/shared per user

Pro Tip: Your NSM may evolve as your product matures. Revisit it quarterly to ensure alignment.

2. Supporting Metrics: The Growth Stack

While your NSM is the compass, supporting metrics are the map. These help you diagnose what's working and what's not across the user journey.

A. Acquisition

- ❖ Cost Per Install (CPI)

- ❖ Click-Through Rate (CTR)
- ❖ Install-to-Signup Rate
- ❖ Top Acquisition Channels

B. Activation

- ❖ Time to First Value (TTFV)
- ❖ Onboarding Completion Rate
- ❖ Activation Rate (e.g., % of users who complete a key action)

C. Engagement

- ❖ Daily/Weekly/Monthly Active Users (DAU/WAU/MAU)
- ❖ Session Length and Frequency
- ❖ Feature Adoption Rates
- ❖ Stickiness Ratio (DAU/MAU)

D. Retention

- ❖ Day 1, Day 7, Day 30 Retention
- ❖ Churn Rate
- ❖ Cohort Retention Analysis

E. Monetization

- ❖ Average Revenue Per User (ARPU)
- ❖ Customer Lifetime Value (CLTV)
- ❖ Conversion Rate (Free → Paid)
- ❖ Revenue by Segment or Channel

F. Referral and Virality

- ❖ Net Promoter Score (NPS)
- ❖ Referral Rate
- ❖ K-Factor (how many new users each user brings)

3. Building a Growth Dashboard

A growth dashboard is your real-time command center. It helps you monitor performance, spot trends, and make faster, smarter decisions.

What to Include:

- ❖ Your North Star Metric
- ❖ Key funnel metrics (acquisition → activation → retention → revenue)
- ❖ Leading indicators (e.g., trial starts, feature usage)
- ❖ Lagging indicators (e.g., MRR, churn)
- ❖ Anomaly alerts (e.g., sudden drop in retention)

Tools to Build With:

Tool	Use Case
Mix panel	Product analytics, funnels, retention
Amplitude	Behavioral cohorts, user journeys
Looker / Tableau	Custom dashboards, SQL-based reporting
Google Analytics 4	Web + app traffic, attribution
Segment	Data pipeline for unified tracking
Databox / Gecko board	Visual dashboards for teams

Tip: Don't just track act. Set weekly rituals to review metrics, identify blockers, and prioritize experiments.

4. Creating a Metrics-Driven Culture

Data is only powerful when it's shared, discussed, and acted upon. Here's how to build a culture where metrics drive momentum:

❖ Make dashboards accessible to all teams not just analysts.

❖ Set OKRs (Objectives and Key Results) tied to product and growth metrics.

❖ Celebrate wins tied to metrics (e.g., "We hit 40% onboarding completion!").

❖ Run retrospectives on failed experiments what did we learn?

Strategic Insight: Metrics should empower, not intimidate. Use them to tell stories, not just report numbers.

5. Avoiding Metrics Missteps

❖ Tracking too many metrics: Leads to noise and confusion.

- Instead: Focus on 1 NSM + 5–7 supporting KPIs.

❖ Chasing vanity metrics: Downloads ≠ success.

- Instead: Prioritize metrics tied to retention, revenue, and user outcomes.

❖ Ignoring qualitative data: Numbers don't tell the whole story.

- Instead: Pair metrics with user interviews, reviews, and support tickets.

❖ Lagging on instrumentation: You can't improve what you don't measure.

- Instead: Implement tracking early even in MVP.

Chapter Takeaway

Metrics are your mobile app's heartbeat. They reveal what's working, what's not, and where to go next. By defining a clear North Star Metric, tracking the right supporting KPIs, and building a culture of data-driven iteration, you empower your team to scale with clarity and confidence.

CONCLUSION
FROM STRATEGY TO EXECUTION: YOUR ROADMAP TO SUSTAINABLE APP GROWTH

"Success is not the result of spontaneous combustion. You must set yourself on fire." **Arnold H. Glasow**

You've just completed a comprehensive journey through the lifecycle of mobile app marketing from market research and brand positioning to launch execution, monetization, and post-launch optimization. Whether you're a founder, product marketer, or growth strategist, you now hold a playbook built for real-world traction and long-term scalability.

But knowledge alone isn't enough. The next step is action deliberate, data-informed, and user-centered.

Recap of Strategic Pillars

Let's revisit the core pillars that will guide your mobile app's success:

1. Market Intelligence

Understand your category, competitors, and customers. Use data to define your ideal customer profile (ICP), identify whitespace opportunities, and position your app with clarity and confidence.

2. Pre-Launch Strategy

Craft a compelling brand identity. Build anticipation through teaser campaigns, influencer outreach, and early access programs. Validate your messaging and UX before launch.

3. Launch Execution

Optimize your app store presence. Leverage paid and organic acquisition channels. Encourage early reviews and activate your referral engine. Launch is not a moment it's a campaign.

4. Post-Launch Growth

Drive retention through personalization, lifecycle messaging, and gamification. Use behavioral data to iterate, expand, and deepen engagement. Build a feedback loop that fuels your roadmap.

5. Monetization

Choose a revenue model that aligns with your value delivery. Test pricing, optimize conversion flows, and balance monetization with user experience. Revenue is a byproduct of relevance.

6. Feedback and Iteration

Treat feedback as a strategic asset. Use surveys, reviews, and usage analytics to prioritize improvements. Close the loop with users and build a culture of continuous learning.

7. Funding and Scale

Match your funding strategy to your growth stage. Whether bootstrapped or VC-backed, focus on traction, efficiency, and storytelling. Scale through product-led growth, partnerships, and global expansion.

Action Plan: From Insight to Implementation

Here's a practical, step-by-step action plan to help you operationalize what you've learned:

Step 1: Define Your Strategic Objectives

* ❖ What does success look like in the next 90 days?
* ❖ Are you optimizing for acquisition, activation, retention, or revenue?
* ❖ Set SMART goals (Specific, Measurable, Achievable, Relevant, Time-bound).

Step 2: Audit Your Current State

❖ Review your app's analytics, user feedback, and competitive positioning.

❖ Identify gaps in onboarding, messaging, monetization, or retention.

❖ Prioritize quick wins and high-impact opportunities.

Step 3: Build a Cross-Functional Growth Plan

❖ Align product, marketing, and customer success around shared KPIs.

❖ Create a roadmap that includes experiments, campaigns, and feature releases.

❖ Assign owners, set timelines, and define success metrics.

Step 4: Implement and Measure

❖ Launch your initiatives with clear tracking in place.

❖ Use dashboards to monitor performance in real time.

❖ Run weekly growth reviews to assess progress and unblock execution.

Step 5: Iterate and Scale

❖ Double down on what works. Kill what doesn't.

- Use cohort analysis, A/B testing, and user interviews to refine your approach.
- Prepare for scale by investing in automation, partnerships, and platform integrations.

Resources for Continued Growth

To stay sharp and ahead of the curve, invest in continuous learning:

Courses: Reforge (for growth), Maven (for product), and Coursera (for mobile UX)

Blogs: Lenny's Newsletter, App Masters, Mobile Dev Memo

Podcasts: Masters of Scale, How I Built This, The Mobile User Acquisition Show

Communities: Indie Hackers, Product Marketing Alliance, Mobile Growth Stack Slack

Events: MAU Vegas, App Growth Summit, SaaStr Annual, Product-Led Summit

Final Words: Your App, Your Impact

The mobile app landscape is dynamic, competitive, and full of opportunity. The strategies in this guide are not static rules they're living systems. Your job is to adapt them, test them, and evolve them in the context of your users, your market, and your mission.

You now have the frameworks, tools, and mindset to:

- ❖ Launch with clarity
- ❖ Grow with intention
- ❖ Monetize with integrity
- ❖ Iterate with humility
- ❖ Scale with confidence

So, take a breath. Reflect on how far you've come. Then take your next step bold, informed, and ready.

Here's to building something remarkable.

GLOSSARY OF TERMS

A/B Testing

A method of comparing two versions of a feature, message, or design to determine which performs better. Users are randomly assigned to each version, and results are measured based on predefined metrics such as conversion rate or engagement.

Acquisition Channels

The platforms, tactics, or sources through which new users discover and install your app. Examples include social media ads, app store search, influencer marketing, and referral programs.

Activation

The moment when a user first experiences the core value of your app. Activation is often tied to a specific action (e.g., completing onboarding, creating a project, sending a message).

Ad Network

A platform that connects app developers with advertisers, enabling the placement of ads inside mobile apps. Examples include Google Ad Mob, Unity Ads, and Meta Audience Network.

App Store Listing

The public-facing page of your app in the Apple App Store or Google Play Store. It includes your title, description, screenshots, icon, ratings, and reviews all of which influence conversions and discoverability.

App Store Optimization (ASO)

The process of improving your app's visibility and conversion rate in app stores through keyword optimization, compelling visuals, and strong ratings and reviews.

Average Revenue Per User (ARPU)

A monetization metric that calculates the average amount of revenue generated per active user over a specific period.

Behavioral Triggers

Automated messages or actions triggered by user behavior, such as inactivity, milestone completion, or hitting usage limits. Used to drive engagement and retention.

Churn

The percentage of users who stop using your app within a given time period. High churn indicates poor retention or product-market misalignment.

Cohort Analysis

A method of grouping users based on shared characteristics (e.g., signup date, acquisition channel) to analyze behavior, retention, and performance over time.

Conversion Rate

The percentage of users who complete a desired action, such as installing the app, subscribing to a plan, or making an in-app purchase.

Customer Acquisition Cost (CAC)

The total cost of acquiring a new user, including marketing, advertising, and sales expenses. CAC is often compared to LTV to assess profitability.

Customer Lifetime Value (CLTV or LTV)

The total revenue a user is expected to generate over the entire duration of their relationship with your app.

Daily/Weekly/Monthly Active Users (DAU/WAU/MAU)

Metrics that measure how many unique users engage with your app daily, weekly, or monthly. These numbers indicate engagement and product stickiness.

Engagement

The depth and frequency of user interaction with your app. High engagement often correlates with strong retention and monetization.

Feature Adoption

The rate at which users begin using a specific feature. Helps identify which features deliver value and which may require redesign or better onboarding.

Freemium Model

A monetization strategy where the app is free to use, but premium features or content require payment. Common in SaaS and consumer apps.

Gamification

The use of game-like elements such as badges, points, streaks, or leaderboards to increase motivation, engagement, and retention.

Go-To-Market (GTM) Strategy

A strategic plan outlining how your app will reach its target audience, acquire users, and achieve revenue goals. Includes positioning, pricing, channels, and messaging.

Growth Loop

A self-reinforcing cycle where user actions generate more users or value. Example: a referral program where each new user invites additional users.

In-App Messaging

Messages delivered inside the app to guide users, announce features, or encourage specific actions. Often used for onboarding, upsells, and engagement.

Iterative Improvement

A continuous process of enhancing your app based on data, user feedback, and experimentation. Central to agile product development.

Key Performance Indicators (KPIs)

Quantifiable metrics used to evaluate the success of your app's marketing, product, or business performance. Examples include retention rate, ARPU, and activation rate.

Minimum Viable Product (MVP)

The simplest version of your app that delivers core value and allows you to test assumptions with real users before investing in full development.

Monetization Model

The method your app uses to generate revenue, such as subscriptions, in-app purchases, ads, or usage-based pricing.

Net Promoter Score (NPS)

A metric that measures user loyalty by asking how likely they are to recommend your app to others. Scores range from −100 to +100.

North Star Metric (NSM)

The single most important metric that reflects the core value your app delivers. It guides product and growth decisions across the company.

Onboarding

The guided experience that helps new users understand your app and reach their first moment of value. Strong onboarding improves activation and retention.

Organic Acquisition

Users who discover your app without paid advertising typically through app store search, word of mouth, or content marketing.

Product-Led Growth (PLG)

A growth strategy where the product itself drives acquisition, activation, and expansion. Users experience value before committing to payment.

Product-Qualified Lead (PQL)

A user who has demonstrated strong intent to purchase based on in-app behavior (e.g., inviting teammates, hitting usage limits).

Referral Program

A system that incentivizes users to invite others to your app, often through rewards, credits, or premium access.

Retention

The ability of your app to keep users returning over time. Strong retention is a key indicator of product-market fit.

Search Engine Optimization (SEO)

The practice of improving your website's visibility in search engines to drive organic traffic. Often complements ASO.

Segmentation

Dividing users into groups based on behavior, demographics, or needs to deliver personalized experiences and targeted marketing.

Session Length

The amount of time a user spends in your app during a single visit. Helps measure engagement quality.

Stickiness Ratio (DAU/MAU)

A metric that shows how often monthly users return daily. Higher stickiness indicates stronger habit formation.

User Flow

The path users take to complete tasks within your app. Optimizing user flows improves usability and conversion.

User Feedback Loop

A structured process for collecting, analyzing, and acting on user feedback to improve the product.

User Interface (UI)

The visual layout and interactive elements of your app, including buttons, icons, colors, and typography.

User Experience (UX)

The overall experience a user has while interacting with your app, including usability, satisfaction, and emotional response.

Virality

The rate at which users share your app with others, leading to exponential organic growth.

REFERENCES

Ailie K.Y. Tang. (2016, October). *Mobile App Monetization: App Business Models in the Digital Era.* International Journal of Innovation, Management and Technology, 7(5). Retrieved from http://www.ijimt.org/vol7/677-MB00017.pdf

Amy Jo Kim. (2018, May 31). *Game Thinking: Innovate smarter & drive deep engagement with design techniques from hit games.* GameThinking.io.

Anastasia Khomych. (2023, March 6) *App Marketing Guide 2021: Best Strategies to Promote Your Mobile App. Get Social.* Retrieved from https://blog.getsocial.im/app-marketing-guide-2020-best-strategies-to-promote-your-mobile-app/

Bank My Cell. *How many smartphones are in the world?* Retrieved from https://www.bankmycell.com/blog/how-many-phones-are-in-the-world

Build Fire. *Mobile App Marketing Costs: How to Plan Your App Marketing Budget.* Retrieved from https://buildfire.com/app-marketing-budget/

Buzinga. *The Advanced Guide to Mobile App Marketing.* Retrieved from https://www.mobuzz.org/wp-content/uploads/2016/11/The-Advanced-Guide-To-Mobile-App-Marketing.pdf

Camilla Koljonen. (2016). *Marketing Plan for a Mobile Application.* Oulu University of Applied Sciences. Retrieved

from
https://www.theseus.fi/bitstream/handle/10024/115326/ThesisC
amillaKoljonen.pdf

Combo App. (2023). *The Ultimate Guide to Mobile App Marketing: 30 Best Tips for 2022.* Retrieved from *https://comboapp.com/services/marketing/mobile-app-marketing-ultimate-guide*

Eric Benjamin Seufert. (2014, February 10). *Freemium Economics: Leveraging Analytics and User Segmentation to Drive Revenue.* Morgan Kaufmann.

Ericsson. *Mobile data traffic outlook.* Retrieved from https://www.ericsson.com/en/reports-and-papers/mobility-report/dataforecasts/mobile-traffic-forecast

Frederick O'Brien. (2020, August 14). *A Smashing Guide to the World of Search Engine Optimization. Smashing Magazine.* Retrieved from https://www.smashingmagazine.com/smashing-guide-search-engine-optimization/

Gary Vaynerchuk. (2018). *Crushing It! How Great Entrepreneurs Build Their Business and Influence—and How You Can, too.* Harper Business.

Google. (2015, May). *Mobile App Marketing Insights: How Consumers Really Find and Use Your Apps.* Retrieved from https://think.storage.googleapis.com/docs/mobile-app-marketing-insights.pdf

L. Ceci. (2022, June 14). *Average time spent daily on a smartphone in the United States 2021.* Statista. Retrieved from https://www.statista.com/statistics/1224510/time-spent-per-day-on-smartphone-us/

Mansoor Iqbal. (2023, May 2). *App Download Data (2023).* *Business of Apps.* Retrieved from *https://www.businessofapps.com/data/app-statistics/*

Mihovil Grguric. (2023, March 7). *Mobile Marketing Trends for 2023 (Including Statements from Industry Leaders).* Udonis. Retrieved from https://www.blog.udonis.co/mobile-marketing/mobile-marketing-trends-in-2020

Neil Patel. *The Definitive Guide to Mobile App Marketing.* Retrieved from https://neilpatel.com/blog/inbound-app-marketing-guide/

Rand Fishkin. (2018, April 24). *Lost and Founder: A Painfully Honest Field Guide to the Startup World.* Portfolio.

S. Dixon. (2022, January). *Daily Time Spent on Social Networking by Internet Users Worldwide from 2012 to 2022.* Statista. Retrieved from https://www.statista.com/statistics/433871/daily-social-media-usage-worldwide/

Tim Maytom. *65 Percent of App Downloads Come from Organic Searches.* Mobile Marketing Magazine. Retrieved from https://mobilemarketingmagazine.com/65-per-cent-of-app-downloads-come-from-organic-searches-app-store-optimisation-adjust-app-annie

Unnati Narang & Venkatesh Shankar. (2019, September). *Mobile Marketing 2.0: State of the Art and Research Agenda.*

Venkata N. Inukollu, Divya D. Keshamoni, Taeghyun Kang, & Manikanta Inukollu. (2014, October 16). *Factors Influencing Quality of Mobile Apps: Role of Mobile App*

Development Life Cycle. Retrieved from
https://arxiv.org/abs/1410.4537

Brian Wong. (2016, September 6). *The Cheat Code: Going Off Script to Get More, Go Faster, and Shortcut Your Way to Success.* Currency.

Julie Zhuo. (2019, March 19). *The Making of a Manager: What to Do When Everyone Looks to You.*

Wes Bush. *Product-Led Growth: How to Build a Product That Sells Itself.* Podcast (October 09, 2024). Retrieved from https://roguestartups.com/episodes/rs330-product-led-playbook-with-wes-bush

George Berkowski. *How to Build a Billion Dollar App: Discover the Secrets of the Most Successful Entrepreneurs of Our Time.* Little, Brown Book Group. Retrieved from https://www.goodreads.com/book/show/23658963-how-to-build-a-billion-dollar-app

ABOUT THE AUTHOR

Nadine Nana Tangpi is a Chief Marketing Officer (CMO), startup growth advisor, and strategic operator with more than 18 years of experience architecting and leading high-impact marketing initiatives. Her career spans dynamic, fast-paced environments where she has consistently helped organizations navigate complexity, overcome growth challenges, and achieve measurable business outcomes.

As a CMO and startup advisor at JumpStart Ventures, Nadine provides hands-on leadership in sales strategy, go-to-market execution, and growth operations. She partners directly with early-stage founders, helping them refine their positioning, accelerate customer acquisition, and build scalable systems that support long-term success. JumpStart Ventures specializes in empowering startups with tailored guidance, and Nadine plays a pivotal role in shaping the trajectory of the companies she supports.

With deep expertise in the startup ecosystem, Nadine understands the realities and pressures that emerging businesses face. She leverages her background in strategic marketing, economics, and communication to help founders craft compelling value propositions, build strong market presence, and execute with clarity and confidence.

Nadine holds an Executive MBA from the Quantic School of Business and Technology, a Master's in Applied Economics from Tuskegee University, and a Bachelor's in Mass Communication from the University of Yaoundé, Cameroon.

Your feedback on this guide is always welcome. Nadine is committed to supporting entrepreneurs, creators, and innovators on their journey toward growth and success. Connect with her to explore how you can collaborate to bring your vision to life.

www.ingramcontent.com/pod-product-compliance
Lightning Source LLC
Chambersburg PA
CBHW070938210326
41520CB00021B/6958